I0493126

TO THOSE IN SEARCH OF
THE TRUTH

TO GENERATIONS OF
CIVILIZATION

Copyright © 2015-2016 Virtumanity Inc. All Rights Reserved.

XU, ChongWei
January 5th, 2016 USA

Keywords: Spacetime topology, General theory of fields and particles, Quantum mechanics and field theories, Thermodynamics, Electromagnetism, General relativity and Gravitation.

PACS: 04.20.Gz, 11.00.00, 03.00.00, 05.70.-a, 41.00.00, 04.00.0

Biography

During 2009-2010, ChongWei XU received a set of "The book of Worlds in Universe" in the old classic manuscripts. Initially the essences of the books appeared as the profound topology of universe in philosophy, until it has concisely revealed the whole scientific structures of Elementary Particles on the Christmas of 2013. Since then, he has been demonstrated the groundbreaking  theories in Particle Physics, Unified Field Theory, on the Origin of Physical States, and beyond.

His recent scientific publications were: "The Christmas gifts of 2013 - Revealing Intrinsic Secrets of Elementary Particles Beyond Quantum Physics" (January 2014), "Theory of Physical Cosmology - Universe Particles" (May 2015), "Theory of YinYang Physics - Horizon Fields" (February 2016), "Unified Theory for All Physics and Beyond" (March 2016), "Unified Field Theory" (May 2016), and the others.

His focus on dialectical nature of philosophy and sciences is to uncover topological framework of the universe, to concisely develop a full intrinsic structure of elementary particles, to philosophically derive the mathematical principles of quantum mechanics, to hierarchically present the unified field theory using a horizon topology, and to heuristically demonstrate the origin of physical states.

Besides, C. Wei Xu created the commercial IPSEC in December 1994, known as the first VPN product in Information Technology histroy. Today, he serves as a chief architect in networking, security, servers, and applications at metropolitan area of Washington, DC.

ChongWei holds the BS and first MS degrees in theoretical physics from Ocean University of China and Tongji University, and the second Master Degree in electrical and computer engineering from University of Massachusetts, USA.

UNIFIED FIELD THEORY

【VIRTUMANITY】 SCIENTIFIC SERIES, VOLUME II

Sciences in Unified Virtual and Physical Reality

Overview

Abstract: The spacetime topology associated with the state conservation of energy and entropy are advanced as extensions beyond the virtual and physical dimensions and curvatures. The spacetime dynamics of manifold continuum, therefore, is uncovered by the natural laws of our universe systematically, philosophically and mathematically, which convey the principles of spacetime movement governing all physical events, transforming universe particles, and constituting extendable physical hierarchies. It develops the dualities of dynamic fields, that form and give rise to physical horizons: from inception of time, energy, mass, and space, to event states, to quantum fields, to thermodynamics, to electromagnetism, gravitational force, and beyond.

The application of an evolutionary process to contemporary theoretical physics therefore derives a complete picture of the principal equations, important assumptions, and essential laws. It prompts the entire discipline of physics, from Newtonian to spacetime relativity to quantum mechanics, to look back to the future: dialectical nature of virtual and physical reality.

Intuitively following the system of spacetime philosophy, this concise theory is accessible and replicable by readers with a basic background in mathematical derivation and theoretical physics.

Table of Contents

Dedications   i

Biography   iii

Overview   vii

Table of Contents   viii

Foreword   x

1. Topology   1

     Generations   1

     Manifolds   2

     Duality   4

2. Infrastructure   7

     Terminology   7

     Framework   10

     Relativity   13

     Fields   16

     Summary   17

3. Conservations   19

     Energy and Entropy Densities   19

     Dynamic Fields   21

     Summary   23

4. Horizon of Quantum Fields   25

    Space Dynamics   25

    Time Dynamics   26

    Summary   29

5. Horizon of Macroscopic Density   31

    Macroscopic Density   31

    Continuity of Virtual Density   32

    Continuity of Physical Density   33

    Summary   34

6. Horizon of Thermodynamics   37

    Bulk Statistics   37

    Thermodynamics   39

    Entropy Extrema   40

7. Horizon of Electromagnetic Fields   43

    Electromagnetic Fields   44

    Covariant Expressions   46

    Physical Potential Dynamics   47

    Summary   48

8. Horizon of Gravitational Fields   49

    Gravitational Fields   49

    Virtual Potential Dynamics   51

    General Relativity   53

    Summary   56

9. Conclusions   57

References   59

Foreword

Everywhere our world shines with a beautiful nature. In every fraction of every creature, we shall find the principles and laws of physics, biology, metaphysics, information technology, and all other sciences. Nature is systematically composed of building blocks, dualities, which take on an abstract form as simple as Space and Time, and as simple as Virtual and Physical existence. Our ancestors discovered that duality orchestrated and harmonized their reality: sun-moon, warm-cold, materialization-consciousness, body-mind, male-female, thought-action, and more. What promise hides in the dualities of physics: space-time, wave-particle, energy-mass, spin-charge, positive-negative, and symmetry-antisymmetry? These dualities are balanced, interdependent, and inexorable. They are manifest in each particular action and movement, the outcome of a dialectical struggle for superiority. The serious study of an honest scientist spends itself on understanding the universe that stands at the very core of our lives. It is essential to believe that the true framework of our universe is a topological hierarchy of virtual and physical duality, flourishing everywhere among the great streams of life, inspiration, and enlightenment.

Following the obvious events of our nature topology, this book systematically unfolds the natural and topological framework as they give rise to remarkable formulae of all principles, laws, assumptions, and operators of classical and modern physics including, but not limited to,

Newtonian Mechanics, Gravity, Electromagnetism, Thermodynamics, Quantum Fields, and universe Manifolds.

Today, our mankind is at dawn of a new era, towards revolutions of:

1. Advancing scientific philosophies towards next generation,

2. Standardizing topological framework for modern physics,

3. Virtualizing informational sciences towards virtue reality,

4. Theorizing biology and biophysics for the life sciences,

5. Rationalizing metaphysics back on the scientific rails.

Our challenge, however, is now greater than that of the trial of Galileo Galilei. Our challenges are not because we lack a profound philosophy of science and an intrinsic theory. Our challenge is, in fact, to leave behind the ambiguous philosophy that we were born with. Our challenge is to open up our minds to facts hidden in the fabric of daily life. Our challenge is to soften our metaphysical prejudices, for the assumption that there is no metaphysical reality is also a metaphysics itself. Our challenge is to overcome all the ignominious desensitized by the clamor of the hype of the world. These are the challenges that we face today, nothing more, but also nothing less.

Xu, Chong Wei   January 5th, 2016
wxu@virtumanity.com
Metropolitan Area of Washington DC, USA

* Virtumanity *
The sciences investigating dialectical nature of virtual and physical reality

CHAPTER 1

Topology

A hierarchical theory is a philosophically and strategically unified formalism aligning with our nature structures such that the mathematics turns out to have close analogues in topology, logic and computation. Our universe has nature objects and structural morphisms, representing events and processes among situations. In modern physics, the nature objects are often virtual and physical matters, and the morphisms are dualities of the dialectical processes orchestrating a set of states in one regime rising and transforming into states of the others: universe topology of the nature structure.

Generations

The terminology of *Space* and *Time* has been in currency since the inception of physics. Throughout the first generation of physics, space and time are individual parameters that have no interwoven relationship. From Newtonian mechanics to Euclidean space[1], the scientific approach known as classical physics seeks to discover a set of physical laws that mathematically describe the motion of bodies under the influence of a system of forces. In classical physics, it is reasonable to interpret a space as consisting of three dimensions, and time as a separate dimension. This regime has presented us with a basic concept for the *Real Existence*

---

of space and the *Virtual Existence* of time, although the virtual existence is hardly studied and their relationship remains unexpressed.

As the second generation, modern physics couples the virtual existence of time with the real existence of space into a single interwoven continuum, known as Spacetime. By combining space and time into a manifold called Minkowski space[2], physicists have significantly simplified a large number of physical theories, as well as described in a more uniform way the workings of the universe at both the supra-galactic and subatomic levels. By revealing their interwoven inferences for the events of a hierarchical universe, the manifold continuum presents us with the enhanced logic for a complex vector of the *Real* dimension of space

$$\mathbf{r} = \{x, y, z\} \equiv \{x_1, x_2, x_3\} \qquad (1.1)$$

and the *Virtual* dimension of time

$$\mathbf{v} = \{ict, \dots\} \equiv \{x_0, \dots\} \qquad (1.2)$$

where the constant c is the speed of light, and *i* marks the virtual or imaginary in mathematics. Although the virtual and real interwoven relationship for dynamics is only limited to physical existence with spacetime curvature, Einstein, one of the greatest minds of the twentieth century, successfully intuited his well-known theory of relativity without using the interwoven continuum of quantum state, fields and energy.

Today, with the acceptance of quantum mechanics, contemporary physics has reached consensus on the possibility of a virtual existence beyond physical reality. When Heisenberg's "uncertainty principle" delimited the duality region of non-physical essence, Bohr emphatically declared that "everything we call real is made of things that cannot be regarded as real."

Manifolds

As proven by the contemporary physics, *Real Space* and *Virtual Time* have a global environment, $G(\lambda)$, composed of events, λ, and constituted by hierarchical structures of both massless objects in virtual dimension, $V(x^\alpha)$, and massive matters in real dimension, $P(x^i)$.

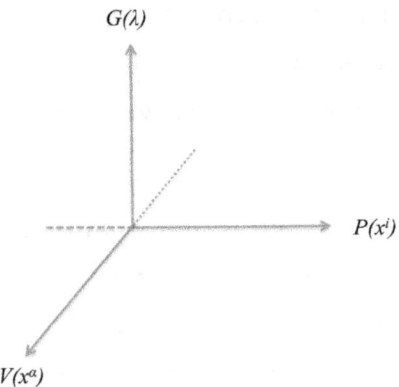

Figure 1.1: Coordinate System of Universe Manifolds

A curve in two-dimensional manifold (the G and P axes, or the G and V axes) is called a *World Line*, corresponding to the history or future of either virtual or physical world but not both. Some examples are as the following:

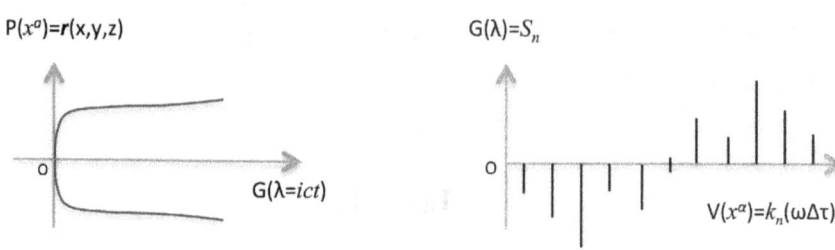

a. Physical Timespace   b. Virtual Statetime

$P(x^a)=r(x,y,z)$ $G(\lambda)=S_n$

$G(\lambda=ict)$ $V(x^a)=k_n(\omega\Delta\tau)$

Figure 1.2 Examples of Physical and Virtual World Lines

As a manifold, the hierarchical structures in the Figure 1.1 − 1.2 are respectively characterized by the real and virtual dimensions of equation (1.1) and (1.2). Together, they localizes a manifold region with common ground as universal conservations: duality $\{v, r\}$. A duality is a set of virtual and real dimensions labeled as a complex vector of x^μ.

$$x^\mu \in \{\mathbf{v}, \mathbf{r}\} = \{x_\alpha, x_i\} \tag{1.3}$$

The space and time duality is the complementary opposition of inseparable, reciprocal pairs of all natural states and events $G(\lambda)$. That, therefore, forms the manifold continuum $\Gamma(x^\mu)$, called *Spacetime Curvature,* shown by the following coordinate system:

$$ds^2 = d\mathbf{v} \cdot d\mathbf{v} + d\mathbf{r} \cdot d\mathbf{r} \tag{1.4}$$

$$ds^2 = g_\nu^\mu dx^\mu dx^\nu \quad : g_\nu^\mu = \delta_{\mu\nu} = \begin{cases} 1 & (\mu = \nu) \\ 0 & (\mu \neq \nu) \end{cases} \tag{1.5}$$

where s is a curvature length or spacetime interval, δ_ν^μ is *Kronecker* delta, g_ν^μ is a simple metric signature $I^+=(+,...,+,+,+)$, and the summation convention is the implicit sum of identical upper and lower indices over all their possible values. Latin indices are only used for the spatial components.

Duality

Spacetime curvature is either virtual whenever $ds^2 < 0$ or real whenever $ds^2 > 0$, respectively, called the *Virtual* or *Real Primacy* of spacetime manifold. The concept of virtual and real spacetime continuum presents the two-sidedness of any event, space and time, each dissolving into the other in an alternating stream that generates a life of situations, movements, or actions through continuous helix-circulations in a universe manifold.

CHAPTER 2

Infrastructure

The fundamental concept of topology *of* the Universe is a holistic approach for the intimate interconnections among the fundamental domains: while the objects in its virtual regime emerge the various transition fields, the states in the scope of *timestate* are composed with both time and energies. This institutes the space reality in *spacetime* domain, from microscopic to macroscopic, composed of various dimensions for relative movements and synergies. Both time and space are the energy spectra, generated by supernaturally interactional fields and their virtual-physical duality.

In physical worlds, they appear in the form of the earthly dimensional space and the heavenly virtual time, affiliating with chronicle events.

Terminology

"Universe" is the whole of everything in existence that operates under a system of topologically-ordered natural laws for, but not limited to, physical matter and virtual objects. Some of its fundamental terminology can be outlined as a preliminary to discussing universe structure.

1. *World - An environment composed of events or constituted by hierarchical structures of massless objects, massive matters, or both. These hierarchical structures are respectively defined as virtual world,*

---

physical world, and universe. Traditionally, for example, the virtual world is referred to as the inner world, the physical world as the outer world, and together they form holistic lives in universe. A world has a permanent form of topology, localizes a region of universe, and interacts with other worlds rising from one or the others with common ground in universal conservations. Furthermore, there are multiple levels of inner worlds and outer worlds. Inner worlds are instances of situations, with or without energy or mass formation, while outer worlds include physical mass of living beings and inanimate objects. Both are real, as well as topologically interactive with and external to each other.

2. *Manifold - A common global environment determined by projections from objects of virtual world. Manifold manifests as various states called global domains, which, for example, emerge as virtual objects of messaons[1], time events of transformations, and state energies of enclaves. They form the ground foundation for the physical reality known as elementary particles with signatures of spin, charge and mass. Once in the physical domain, called spacetime, they continuously and progressively rise through various stages of physical formations. Together, each advances from the others under a topological hierarchy of the universe to develop a consistent system of stages. The domain of*

this consistent system of stages is divided into various scopes, called horizon.

3. *Horizon - The apparent boundary of a realm of perception or the like, where unique structures are evolved, topological functions are performed, various neighborhoods form interactions, and worlds are composed through transformations. Each horizon rises and contains specific fields as a construction of the dynamics within or beyond its own range. In other words, fields vary from one horizon to the others, each of which is part of and aligned with the horizon topology of the world.*

4. *Duality - The complementary opponents of inseparable, reciprocal pairs of all natural states and events. Among them, the most fundamental duality is known as Real/Physical "-" and Virtual "+", with neutral balance "o" that appears as if there were nothing. Spacetime presents the two-sidedness of any event or world, each dissolving into the other in an alternating stream that generates the life of situations, movements, or actions through continuous helix-circulations in a universe manifold. Because of this nature, a universe manifold always has a mirrored pair in the imaginary part, a conjugate pair of a complex manifold, defined as* **Spacetime Manifold**.

5. *Light - An event object emanated in virtual word, determined by spacetime phases, and confined by a*

manifold of virtual world called xingspace. In physical world, the speed of light is only variable with time as a function of virtual position, not space, for all observers, regardless of the physical motion of the light source. The speed of light forms a dimension reflected from virtual world, with the property of being confined by spacetime phases in xingspace and of appearing as a universal constant in and only in physical space.

In general, physics studies duality of symmetric-antisymmetric formations, state-energy movements, mass-massless kinetics, and time-space manifolds within a scope of mass-massless environment for the fundamental topology of physical-virtual dynamics.

Framework

Considering the physical formation from particle to cell to organism to life, it is natural to add to the terminology developed to describe the organization of universe up to the level of the particle. Thus, we present two principles of topological frameworks: Transformation and Revelation.

Transformation is a type of the bi-directional process of evolution and stagnation in physical world, which emanates from or conceals in resources of the virtual worlds.

For example, an elementary particle is composed of objects that exist in various forms beyond physical world. Those objects exist in environments of virtual worlds, which may not be directly detectable by measurements in our physical world because of the limitation of uncertainty. However, it is indirectly sensible and appears throughout all physical existence. On occasion, it may even be explicitly formulated, such as the philosophy of spacetime, which was discovered seven thousand years ago.

Originated by supernatural, particle objects, composited by the virtual elements with spacetime appearance, are traveling across multi-worlds that can conduct and perform activities as a part of their behaviors at outer worlds. This is a bi-directional transforming seamlessly between the virtual and physical worlds. In a sense of physics, the transformation between virtual and physical worlds involves the environments of

1. *Spacetime nature as a part of fully-virtual world, named Xingscope or Xingspace manifold;*
2. *Energy enclave as a part of virtual and physical worlds, named Statescope or Timestate manifold;*
3. *Mass embody as a part of fully-physical world, named Spacescope, or Spacetime manifold.*

More details on their formation and the characteristics of the entire elementary particles are available at reference [31].

When virtual objects form the existence as a matter, cosmology of a universe is in coherent harmonization of supernatural evolutions emerging spacetime dualities of virtual reality. The duality is the indivisible whole and exhibited in all physical matters. For physicists, examples of these fundamental instances are a duality of symmetry-antisymmetry, state-energy, time-space, mass-massless, wave-particle, and much more. For metaphysicists, obvious examples are a duality of male-female, body-mind, thought-action, consciousness-brain, and more beyond. They are complementary interdependent, and can manifest balance or supremacy in one against the other to perform particular actions or movements of objects or events based on criterions of the situation.

Revelation is another type of evolution and stagnation processes within a world, ascending from or descending into each of the layers, relatively and respectively.

In physical world, there exists numerous levels of reality in a variety of variations so that one level forms the others that are aligned with its topological hierarchy. Each level of physical reality constitutes a horizon, which contains or yields the following actions:

1. Forms a domain of its principles commonly shared by their own behaviors and activities;
2. Evolves into its higher level of horizons to advance revelations as a natural growth process;
3. Diminishes to its lower level of horizons to recycle resources as an inanimate concealment.

Therefore, within each of horizons, there exists a unique interactions of formations, fields, forces, functions, information, and the messengers, relatively and respectively.

Relativity

Virtual Time is a set of virtual actions performed on a physical system. With fixed spatial coordinates x_i, the elapsed interval as it moves forward in time is negative, $ds^2 = d\mathbf{v}^2 = -c^2 dt^2 < 0$. Along a trajectory through spacetime, this elapsed time, called universe time t_u, will be the actual time measured by the local observers.

$$t_u = \frac{1}{c}\int \sqrt{-ds^2} = \int \sqrt{-\eta_\nu^\mu \frac{dx^\mu}{d\lambda}\frac{dx^\nu}{d\lambda}}\, d\lambda \qquad (2.1)$$

where the four spacetime coordinates $x^\mu(\lambda)$ are a function of global parameter λ.

The virtual time variation interacting with a physical movement can be expressed by its space coordinate vector of position $\mathbf{r}$ and velocity vector of speed $\mathbf{u} \equiv d\mathbf{r}/dt$. For spacetime curvature, relative time t' can be derived as the following characteristics:

$$dt' = \frac{\partial(ds)}{\partial x_o}dt = \frac{dx_o + d\mathbf{r}\cdot d\mathbf{r}/dx_o}{\sqrt{\left(\frac{dx_o}{dt}\right)^2 + \left(\frac{d\mathbf{r}}{dt}\right)^2}} = \frac{dt - d\mathbf{r}\cdot\frac{\mathbf{v}}{c^2}}{\sqrt{1-\left(\frac{\mathbf{v}}{c}\right)^2}}$$

$$\frac{dt'}{dt} = \frac{dx_0'}{dx_0} = \sqrt{1-\left(\frac{\mathbf{u}\cdot\mathbf{u}}{c}\right)^2} \equiv \frac{1}{\gamma} \qquad (2.2)$$

All observers must agree on the same natural principles, disregarding the observer's reference frame.

A general Lorentz transformation L_ν^μ is a linear map from X to X' of the form

$$(X')^\mu = L^\mu_\nu X_\nu \tag{2.3}$$

$$L^T \eta L = \eta \quad \Rightarrow \quad L^\rho_\mu \eta_{\rho\sigma} L^\sigma_\nu = \eta_{\mu\nu} \tag{2.4}$$

where $\eta_{\mu\nu}$ is the spacetime metric of: $\eta_{\mu\nu} = diag(1,1,1,1)$. For a 4-vectors, it has invariant quantities that all observers can agree on, disregard to their reference frame. For example, the square of the distance from the origin to some point in spacetime labelled by X is

$$X^\mu X_\nu = \eta^{\mu\nu} X_\mu X_\nu = x_0^2 + x_1^2 + x_2^2 + x_3^2 \qquad : x_0 \equiv ict \tag{2.5}$$

They are the invariant intervals for time-space or thermo-space, respectively.

The mathematics lying behind are the two types of coordinates of X^μ and X_μ formally representing in different notations of the observed spaces. To distinguish them, we refer to X^μ as vectors and call X_μ covectors. The components of the vector X^μ are sometimes said to be contra-variant while the components of the covector X_μ are to be a covariant. They operate as the following illustration:

$$\partial^\mu \partial_\nu = \partial_\nu \partial_\nu = \partial^\nu \partial^\nu = \frac{\partial^2}{\partial x_\alpha^2} + \nabla^2 \quad : \nabla \equiv e_i \frac{\partial}{\partial x_i} \tag{2.6a}$$

$$\partial_\mu \equiv \frac{\partial}{\partial x_\mu}, \quad \partial^\mu \equiv \frac{\partial}{\partial x^\mu} \qquad : x^\alpha, x_\alpha \in \mathbf{v}, \quad x^i, x_i \in \mathbf{r} \tag{2.6b}$$

The solutions to (2.4) fall into a few classes for rotations, and movements as the following:

$$\Lambda_\nu^\mu = \begin{pmatrix} 1 & 0 & 0 & 0 \\ 0 & & & \\ 0 & & R & \\ 0 & & & \end{pmatrix} \qquad \Lambda_\nu^\mu = \begin{pmatrix} \begin{pmatrix} A & -B \\ -B & A \end{pmatrix} & \begin{pmatrix} 0 & 0 \\ 0 & 0 \end{pmatrix} \\ \begin{pmatrix} 0 & 0 \\ 0 & 0 \end{pmatrix} & \begin{pmatrix} 1 & 0 \\ 0 & 1 \end{pmatrix} \end{pmatrix} \qquad (2.7)$$

where the 3×3 rotation matrix R is in the spatial part, obeying $R^T R = 1$, while the 2x2 matrix with the elements of A and B is the motion part. These transformations describe how to relate the coordinates of two observers rotated with or moving relation respect to each other.

Based on the cosmological principle of universe particles[31], virtual time associated with energy first occurs globally, then in physical space, and finally in the inception of mass. Laws of physics are conserved and irrelevant to observational transformations, invariant in all reference systems with either inertial or accelerating frames of observation.

Transformation between reference frames do not change the laws of physics, although it does alter its description relative to an observer. If the observer is not within the set of observed objects, relativity must be considered as an external effect in addition to an "isolated" system. This means that "closed" objects are "open" to relativity for observers. In particular, the reference frame must consider itself as the dynamic cause of its own relative movement.

Fields

Field is a function quantity of an object presenting an interaction value to its surrounding spacetime curvature, arisen by or acting on its opponent through a duality of their spacetime reactions.

Scalar Field is a single magnitude as a variable component, at which each point of scalar is unchanged by an observer's transformation:

$$\Phi'(X') = \Phi(X) \tag{2.8}$$

This is a rank-0 transformation rule, or the 0-form tensor carrying no indices. The scalar product has no tensor indices left and is said to be fully contracted.

Vector Field is a set of variable components, which mimics the transformation law of the scalar field gradient, called covariant vector, or the coordinate differential, called contravariant vector, respectively. The vector product results in a metric tensor indices left and is said to be fully contracted. At every point in spacetime, there are four linearly independent basis vectors e_μ so that a flat spacetime can be denoted by $\{e_0, e_1, e_2, e_3\}$ as the mutually orthogonal unit, called an orthonormal basis. When the derivative operator applies to the vector of space coordinate x^μ, it defines the spacetime movements of the velocity in four-dimensional manifold:

$$\mathbf{U} = U^\mu \mathbf{e}_\mu \qquad \mathbf{u} = \frac{\partial \mathbf{r}}{\partial t} = \frac{\partial x_1}{\partial t}\mathbf{e}_1 + \frac{\partial x_2}{\partial t}\mathbf{e}_2 + \frac{\partial x_3}{\partial t}\mathbf{e}_3 \tag{2.9a}$$

$$U^\mu \in ic\frac{\partial^\mu}{\partial x_0}x^\mu = (ic, \mathbf{u}), \quad U'^\mu \in \frac{\partial x_0}{\partial x'_0}U^\mu = \gamma(ic, \mathbf{u}) \tag{2.9b}$$

$$g_{\mu\nu}U^{\mu}U^{\nu} = c^2 \tag{2.9c}$$

where U^{μ} or U'^{μ} is a covariant vector, which is a complex vector system as a duality of virtual and physical manifold.

In the geometric context of four-dimensional spacetime, the four-current density can be defined as

$$J^{\nu} \in \left(ic\rho, \mathbf{J}\right) \qquad : \mathbf{J} \in (J_1, J_2, J_3) \tag{2.10}$$

where ρ is the charge density and $\mathbf{J}$ the conventional and special current density. Similarly, the four-momentum and four-forces are defined as the following:

$$\mathbf{P}' = m_0\mathbf{U}' = m_0U'^{\mu} \in \gamma(iE/c, \mathbf{p}) \qquad\qquad : \mathbf{p} \in (p_1, p_2, p_3) \tag{2.11}$$

$$\mathbf{F}' = ic\frac{d\mathbf{P}}{dx_0} = \gamma(\frac{i}{c}\frac{dE}{dt}, \frac{d\mathbf{p}}{dt}) = \gamma(\frac{i}{c}\mathbf{f}\cdot\mathbf{u}, \mathbf{f}) \quad : \mathbf{f} \in (f_1, f_2, f_3) \tag{2.12}$$

where m_0 is the rest or macro mass of an object, $\mathbf{p}$ and $\mathbf{f}$ are the conventional and special momentum and force, respectively.

Summary

Under the universe topology, the terminology and frameworks are philosophically constructed and represented as a foundation of the infrastructure, upon which the mathematical methods are outlined as the basic tools for discovering our nature in physics. Therefore, it might enable us the guidance to perform concisely in quest of the truth using heuristic, intuitive, and approximate arguments, supporting the application of mathematics to problems in physics and the development of mathematical methods suitable to our nature hierarchy, observations and experiments.

CHAPTER 3

Conservations

In a manifold, spacetime densities of energy and state are the essential properties in the manifold curvature. Energies are mutable in the transformation between virtual $(ds^2 < 0)$ and real $(ds^2 > 0)$ spacetime curvature, between time and space, or between massless and massive objects of a matter. In physical reality, energy emanates during physical inceptions of time evolution and materialization, and appears to be inexorable, intractable, and transferable among various types of interactions. The maintenance of energy and state densities tends to distribute their equilibria between the virtual and real spacetime continuum.

Energy and Entropy Densities

If a manifold system has internal energy density U, spacetime energy density L, kinetic-energy density T, these energy densities yield an equilibrium equation for their state densities to an external observer O:

$$U = L + T \qquad\qquad (3.1)$$

This principle is called *Manifold Conservation of Energy Density*, or simply *Spacetime Energy Conservation*.

Therefore, energy in the complex manifold of spacetime continuum $\Gamma(x^\mu)$ shall have complex fields associated to its state functions. For that, the virtual time and real space fields must exist in accommodation with the complex dimension of Manifold. Traditionally, the total state function is denoted as Ψ while the image of the state function is $\Psi*$. To generalize, rewrite the real and virtual state functions, each in term of the manifold density for spacetime continuum as one of the global environment $G(\Gamma, O)$:

$$\rho_\psi = \Psi^+ (\Gamma, O)\, \Psi^- (\Gamma, O) = L\,(\Gamma, O) \tag{3.2}$$

where the state functions of Ψ^- are the density in physical spacetime, while Ψ^+ are the state density functions in virtual spacetime, where the signs of "-" and "+" indicate the spacetime twin, similar to complex conjugate in mathematics.

In spacetime, entropy is a measure of the specific number of ways in which a manifold system could be arranged towards either order or disorder. When an entropy decreases, the intrinsic order, or development, of virtual spacetime into physical spacetime is more dominant than the reverse process. Conversely, when an entropy increases, the intrinsic disorder, or chaos, becomes dominant and conceals physical resources into virtual spacetime. As a state process, formless spacetime entropy is a scalar function measuring the total change of state density ρ_ψ between energy-density equilibria:

$$S_\psi = -\,k_s \int \rho_\psi d\Gamma \Rightarrow extrema \tag{3.3}$$

where k_s is a constant. Combined with the equations above, and assuming dynamic fields $\phi_n^-(x^\mu)$ of real-space-primacy states, $\phi_n^+(x^\mu)$ of virtual-time-primacy states, and spacetime coordinates x^μ, the equation of state entropy may be rewritten as the following

$$S_\psi = -k_s \int \rho_\psi d\Gamma = -k_s \int L(\phi_n^-, \partial^\mu \phi_n^-, \phi_n^+, \partial^\mu \phi_n^+) d\Gamma \qquad (3.4)$$

The energy equilibrium in form of (3.1) is similar to **Lagrangian**[3] density introduced in 1788, although the new sets of the variables have alternated. This principle is called *Manifold Conservation for Entropy Density* or simply *Spacetime Entropy Conservation*.

Dynamic Fields

At the first horizon, the operator ∂^μ is given with reference to the 4-tuple coordinates, and to the spacetime duality. For space fields of state ϕ_n^-, there are the following equivalent operators:

$$\partial_\alpha \phi_n^- \equiv \frac{\partial}{\partial x_\alpha} \phi_n^-, \quad \partial_i \phi_n^- \equiv \nabla \phi_n^- \qquad : x_\alpha \in \mathbf{v} \equiv \{ict\}, x_i \in \mathbf{r} \qquad (3.5)$$

In duality with time fields of state ϕ_n^+, these operators are associated with its opponent as a "-" sign during spacetime operations, because the ϕ_n^+ is subject to the opposing energies of space states, illustrated as the following:

$$\partial_\alpha \phi_n^+ \equiv -\frac{\partial}{\partial x_\alpha} \phi_n^+, \quad \partial_i \phi_n^+ \equiv \nabla \phi_n^+ \qquad : x_\alpha \in \mathbf{v} \equiv \{ict\}, x_i \in \mathbf{r} \qquad (3.6)$$

As a natural principle, entropy tends towards extrema between the equilibria of the fixed end-states $\delta S_\psi = 0$ that either increases for disorder processes or decreases for order processes, which derives the motion equation of dynamic fields[16]:

$$\partial^\mu \left(\frac{\partial L}{\partial (\partial^\mu \phi)} \right) - \frac{\partial L}{\partial \phi} = 0 : \quad \phi \in \{\phi_n^-, \phi_n^+\} \tag{3.7}$$

known as the **Euler-Lagrange**[2] partial differential equation for the action of a system, introduced in the 1750s.

In the universe manifolds, a function of an inception process can be represented as an infinite sum of terms that are calculated from the values ξ of the function's derivatives at an initiation point ξ_0, shown as the following

$$L\left(\xi\right) = L\left(\xi_0\right) + L'\left(\xi_0\right)(\xi - \xi_0) + \ldots \frac{L^n\left(\xi_0\right)(\xi - \xi_0)^n}{n!} \tag{3.8}$$

known as the **Taylor** and **Maclaurin** series, introduced in 1715[3].

During the spacetime transformation, the timestate density $\phi_n^- \phi_n^+$ is incepted, $\xi_0 = 0$, by virtual time evolution of $\dfrac{\partial}{\partial x_0}$, shown as the following:

$$\xi = -\frac{\partial}{\partial x_0}, \quad \xi_0 = 0 \tag{3.9}$$

Applying equation (3.9) to the equation (3.8), the spacetime energy density L has its transformation process in the form of kinetic energy density T :

$$T = -(\frac{\kappa_\tau}{2}\frac{\partial}{\partial x_0} + \kappa_{\tau2}\frac{\partial^2}{\partial x_0^2} + \ldots)(\phi_n^- \phi_n^+) \tag{3.10}$$

where κ_τ and $\kappa_{\tau2}$ are coefficients of the first and second orders defined as the timestate coefficients.

Considering $U = U_l + U_x$, the density forms the internal "local" energy of U_l with the bulk potential of the system $V(\mathbf{x}, t)$, as the internal energy density :

$$U_l = V(\mathbf{x}, t)\phi_n^- \phi_n^+ \tag{3.11}$$

Meanwhile, it is this "local" density at $\xi_0 = 0$ that emerges as physical potential, $-\nabla\phi^\pm$, of the energy density U_x, shown as the following:

$$\xi = -\nabla\phi^\pm, \qquad \xi_0 = 0 \tag{3.12}$$

Applying equation (3.12) to the equation (3.8), the spacetime energy density L has its transformation process in the form of "local" energy potential U_x:

$$U_x = \kappa_x[\nabla\phi_n^+ + \alpha_2^+ (\nabla\phi_n^+)^2 \cdots][\nabla\phi_n^- + \alpha_2^- (\nabla\phi_n^-)^2 \cdots] \tag{3.13}$$

where κ_x and $\alpha_2^\pm$ are the spacetime coefficients of the first and second orders.

Summary

The spacetime continuum is presented as a topology nature of our universe. The principles convey the law of energy and entropy in state conservations parameterized by the complex dimensions of virtual and physical existence in spacetime curvature, movement and interactions.

With the concepts of spacetime interactions, natural duality, and the equilibria formulations from the extrema equations above, we are now ready to derive horizon fields of Quantum Dynamics in the next section.

CHAPTER 4

Horizon of Quantum Fields

Spacetime presents the two-sidedness of any dynamic fields, each dissolving into the other in an alternating stream of the time dynamics of ϕ_n^+ fields and the space dynamics of ϕ_n^- fields. Together, they generate a duality of motion transform processes.

Space Dynamics

Rising from the time fields of ϕ_n^+ and $\partial_\mu \phi_n^+$, the space dynamic fields represent their field equation (3.1) of the spacetime energy density $L = U_x + U_l - T$ in the terms of its motion transform processes, given by the equations of (3.10), (3.11) and (3.13).

As the consequences, the space dynamic fields give rise to the time motion transform processes approximated at the first and second order of perturbations as the following:

$$\frac{\partial L}{\partial \phi_n^+} = \frac{\kappa_\tau}{2} \frac{\partial \phi_n^-}{\partial x_0} + \kappa_{\tau 2} \frac{\partial^2 \phi_n^-}{\partial x_0^2} + V(\mathbf{x}, t)\phi_n^- \qquad (4.1a)$$

$$\frac{\partial}{\partial x_0}[\partial L / \partial(\frac{\partial \phi_n^+}{\partial x_0})] = -\frac{\kappa_\tau}{2} \frac{\partial \phi_n^-}{\partial x_0} - 2\kappa_{\tau 2} \frac{\partial^2 \phi_n^-}{\partial x_0^2} \qquad (4.1b)$$

$$\nabla(\frac{\partial L}{\partial(\nabla \phi_n^+)}) = \kappa_x(\nabla^2 \phi_n^- + 2\alpha_2^+ \nabla^2 \phi_n^-) \qquad (4.1c)$$

From these relations, the motion equations of *(3.7)* determine the following relationship of physical state functions, illustrating physical dynamics rising from virtual interactions:

$$3\kappa_{\tau 2}\frac{\partial^2 \phi_n^-}{\partial x_0^{\;2}} + \kappa_\tau \frac{\partial \phi_n^-}{\partial x_0} - 2\kappa_x a_2^+ \nabla^2 \phi_n^- + \hat{H}\phi_n^- = 0 \qquad (4.2a)$$

$$\hat{H} \equiv -\kappa_x \nabla^2 + V(x_0, \mathbf{x}) \qquad (4.2b)$$

where $\hat{H}$ is defined as the relationship known as **Hamiltonian**[4], introduced in 1834. For the first order of the internal energy and kinetic-energy, the equation (4.2) emerges as the **Schrödinger** equation[5], introduced in 1926, in the form of:

$$i\hbar\frac{\partial \phi_n^-}{\partial t} = \hat{H}\phi_n^- \quad : \kappa_\tau = \hbar c, \kappa_x = \hbar^2/2m, \kappa_{t2} = a_2^+ = 0 \qquad (4.3)$$

where $\hbar$ is the **Planck** constant[6], introduced in 1900. This represents the statetime dynamics as the space equation of quantum state fields rise from its opponent virtual time fields during the duality of time and space evolutions.

Time Dynamics

Rising from the space fields of ϕ_n^- and $\partial_\mu \phi_n^-$ in parallel fashion, the time dynamic fields represent their spacetime energy density equation of *(3.1)* $L = U_x + U_l - T$ in terms of its duality in motion transform processes, given by the equations of (3.10), (3.11) and (3.13).

Therefore, the duality of time dynamic fields are given risen to the following the motion transform processes approximated at the first and second order of perturbations:

$$\frac{\partial L}{\partial \phi_n^-} = -\frac{\kappa_\tau}{2}\frac{\partial \phi_n^+}{\partial x_0} + \kappa_{\tau 2}\frac{\partial^2 \phi_n^+}{\partial x_0^2} + V(\mathbf{x}, t)\phi_n^+ \qquad (4.4a)$$

$$\frac{\partial}{\partial x_0}[\partial L / \partial(\frac{\partial \phi_n^-}{\partial x_0})] = -\frac{\kappa_\tau}{2}\frac{\partial \phi_n^+}{\partial x_0} + 2\kappa_{\tau 2}\frac{\partial^2 \phi_n^+}{\partial x_0^2} \qquad (4.4b)$$

$$\nabla(\frac{\partial L}{\partial(\nabla \phi_n^-)}) = \kappa_x(\nabla^2 \phi_n^+ + 2\alpha_2^-\nabla^2 \phi_n^+) \qquad (4.4c)$$

From these relations, the motion equations of (3.7) determines another set of linear partial differential equation of time state function, illustrating virtual dynamics rising from its opponents in the physical interactions:

$$-\kappa_{\tau 2}\frac{\partial^2 \phi_n^+}{\partial x_0^2} - 2\kappa_x\alpha_2^-\nabla^2 \phi_n^+ + \hat{H}\phi_n^+ = 0 \qquad (4.5)$$

where $\hat{H}$ is a *Hamiltonian* operator. Associated with the coefficients of $\hbar$ and m, the time fields of quantum dynamics are generalized as to the equation of:

$$\frac{1}{c^2}\frac{\partial^2 \phi_n^+}{\partial t^2} - \nabla^2 \phi_n^+ + \frac{m}{\hbar^2}\hat{H}\phi_n^+ = 0 \quad : \kappa_{\tau 2} = 2\kappa_x\alpha_2^- = \frac{\hbar^2}{m} \qquad (4.6)$$

For a point particle, it can be further represented as:

$$\hat{H}\phi_n^+ = E\phi_n^+ \qquad E = mc^2 \qquad (4.7)$$

where m is the rest mass of a point object, and $E = mc^2$ is known as **Einstein** equation[7], introduced in 1905. As a consequence, the equation (4.5) becomes the following field equation of quantum dynamics:

$$\frac{1}{c^2}\frac{\partial^2 \phi_n^+}{\partial t^2} - \nabla^2 \phi_n^+ + \left(\frac{mc}{\hbar}\right)^2 \phi_n^+ = 0 \qquad (4.8a)$$

known as **Klein–Gordon** equation[9] introduced in 1928. In covariant notation, its 4-dimensional operation is as the following:

$$\left[-\delta_{\mu\nu}\partial_\mu\partial_\nu + \left(\frac{mc}{\hbar}\right)^2\right]\phi_n^+ = 0 \qquad (4.8b)$$

It represents the relativistic field functions of the quantum states under time field in the spacetime domain. This equation also derives the energy-momentum conservation equation of:

$$E^2 = \mathbf{P}^2 + (mc^2)^2 \quad : -i\hbar\nabla \rightarrow \mathbf{P}, \quad -\hbar c\frac{\partial}{\partial x_0} \rightarrow E \qquad (4.9)$$

known as the relativistic equation relating any object's rest or intrinsic mass m with total energy E, and momentum $\mathbf{P}$.

The above field equation implies the spacetime duality of interactions with their space and time operators of $\hat{Y}_t^-$ and $\hat{Y}_t^+$ by the following:

$$\left(\hat{Y}_t^-\hat{Y}_t^+ + 1\right)\phi_n^+ = 0 \quad : \hat{Y}_t^\pm = \frac{\hbar}{mc}\left(i\frac{\partial}{\partial x_0} \pm \nabla\right) \qquad (4.10)$$

The space actions of unit operator, $\hat{Y}_\tau^-$, is a statetime process of virtual reproduction, while the time actions of unity operator, $\hat{Y}_\tau^+$, is a spacetime process of physical annihilation.

Summary

The manifold conservations govern the spacetime events and constitute the dynamics that form and give rise to both state fields of virtual time and physical space for spacetime dynamics. Following the logic, the model derives, but not limited to, the *Schrödinger* and *Klein–Gordon* quantum equations as a manifold duality of real space and virtual time interweavement.

CHAPTER 5

Horizon of Macroscopic Densities

In the spacetime, arising from the internal symmetry and antisymmetry of the first horizon, the spacetime duality forms up the second horizon as the group effects of a conserved current, characterized by the spacetime components of standard coordinates inherent from the first horizon. Associated with the space or time fields of primary state $\phi_n^{\mp}$, the internal nature produces each of opposite dualities as complex conjugate $\varphi_n^{\mp}$, an integrity of which statistically represents the Macroscopic Densities

Macroscopic Density

For an observable value O in a macroscopic system, the ensemble is in a mixed state such that each pair of the primary states $\phi_n^{\mp}$ and the auxiliary states $\varphi_n^{\mp}$ occurs with probability $p_n(T)$ as a horizon factor, where T is temperature in units of **Kelvin**[10], introduced in 1848. Together, they form macro objects $< \hat{O}, \phi_n >$ defined as the following for the observable density of an operator $\hat{O}$.

$$\left\langle \hat{O}, \phi \right\rangle \equiv \sum_n p_n(\varphi_n \hat{O} \phi_n - \phi_n \hat{O} \varphi_n) \quad : \sum_n p_n = 1 \qquad (5.1a)$$

$$\hat{O} \in \{\partial/\partial x_0, \nabla\}, \quad \phi_n \in \{\phi_n^-, \phi_n^+\}, \quad \varphi_n \in \{\varphi_n^-, \varphi_n^+\} \qquad (5.1b)$$

where the horizon factor of $p_n(T)$ raises temperature during formations of a bulk macro-system. In macroscopic horizon, an entity of objects is observable if and only if it satisfies spacetime asymmetry as the following:

$$\varphi_n \hat{O} \phi_n \neq \phi_n \hat{O} \varphi_n \tag{5.2}$$

meaning that the invariance does not follow the symmetric effects of the internal balance of spacetime duality. Therefore, each of the duality in actions has their own dominant primacy in parallel, besides their auxiliary reactions with each other to maintain their common horizon or give rise to the next horizon. With law of the conservation of motion, the above equation illustrates the **Heisenberg** picture[11], introduced in 1925.

Continuity of Virtual Density

Probability of a density ρ_n^+ and a density current of $\mathbf{J}_n^+$ rises from space movements with time field interactions shown by the following integrity:

$$\rho^+ = \sum_n \rho_n^+ = ic \left\langle \partial/\partial x_0, \phi^+ \right\rangle \tag{5.3a}$$

$$\mathbf{J}^+ = \sum_n \mathbf{J}_n^+ = \left\langle \nabla, \phi^+ \right\rangle \tag{5.3b}$$

Since the complex conjugate of state φ_n^+ has similar equations as its opponent state, the time fields of equation (4.6) represent the following equations for both of states:

$$\sum_n P_n \left(\varphi_n^+ \frac{\partial^2 \phi_n^+}{\partial x_0^2} - \varphi_n^+ \nabla^2 \phi_n^+ + \frac{m}{\hbar^2} \varphi_n^+ \hat{H} \phi_n^+ \right) = 0 \tag{5.4a}$$

$$\sum_n P_n \left(\phi_n^+ \frac{\partial^2 \varphi_n^+}{\partial x_0^2} - \phi_n^+ \nabla^2 \varphi_n^+ + \frac{m}{\hbar^2} \phi_n^+ \hat{H} \varphi_n^+ \right) = 0 \qquad (5.4b)$$

Combined with equations (5.3), the time fields of quantum dynamics forms a continuity of density equation that raises the macroscopic horizon:

$$ic \frac{\partial \rho^+}{\partial x_0} - \nabla \cdot \mathbf{J}^+ = -\frac{m}{\hbar^2} \left\langle \hat{H}, \phi^+ \right\rangle \equiv -K_s^+(\hat{H}) \qquad (5.5)$$

where the scaler K_s^+ is the virtual source producing flux continuity, a virtual object of time energy and momentum in spacetime. Because its symmetric entity is cyclic within a point object: $\hat{H}\phi_n^+ = mc^2\phi_n^+$, $\hat{H}\varphi_n^+ = mc^2\varphi_n^+$, the time field appears as if it were not existent, or as if it were physically empty:

$$K_s^+(\hat{H}) = -\frac{m}{\hbar^2} \left\langle \hat{H}, \phi^+ \right\rangle \Rightarrow 0^+ \qquad (5.6)$$

The symbol 0^+ means that, although it is physically undetectable, the time field rises whenever there is a physical object as its opponent in tangible interactions. The equation (5.5) is a time-dependent virtual density, giving rise to a part of the horizon for bulk dynamics.

Continuity of Physical Density

Likewise, from equation of (4.2), there is a space density probability of ρ_n^- and a density current of $\mathbf{J}_n^-$, rising from time field interactions shown as the following integrity of density equations in macroscopic physical movements:

$$3\kappa_{\tau 2}\frac{\partial \rho^-}{\partial x_0} - 2\kappa_x \alpha_2^- \nabla \cdot \mathbf{J}^- = i\frac{\kappa_\tau}{c}\rho^- - \left\langle \hat{H}, \phi^- \right\rangle \qquad (5.7)$$

$$\rho^- = \sum_n \rho_n^- = ic\left\langle \frac{\partial}{\partial x_0}, \phi^- \right\rangle, \quad \mathbf{J} = \sum_n \mathbf{J}_n^- = \langle \nabla, \phi^- \rangle \qquad (5.8)$$

Therefore, the space fields of quantum dynamics forms a continuity of physical density equation that raises the spacetime duality to a new horizon:

$$ic\frac{\partial \rho^-}{\partial x_0} + \nabla \cdot \mathbf{J}^- = -K_s^-(\rho^-, \hat{H}) \quad : \frac{3\kappa_{\tau 2}}{2\kappa_x \alpha_2^+} = -ic \qquad (5.9)$$

$$K_s^-(\rho^-, \hat{H}) = \frac{1}{2\kappa_x \alpha_2^+}\left(\frac{i\kappa_\tau}{c}\rho^- - <\hat{H}, \phi^- >\right) \approx -\frac{<\hat{H}, \phi^- >}{2\kappa_x \alpha_2^+} \qquad (5.10)$$

where the scaler $K_s^-(\rho^-, \hat{H})$ is a resource balanced to its own density ρ^- as its physical existence in the macroscopic horizons. The above equations are physical equations of time-dependent space density, rising into another twin part of the horizon for bulk dynamics.

Summary

The macroscopic effects of dynamics can be statistically derived by the spacetime interactions as density equations of continuity in the form of covariant tensors:

$$\partial_\nu J_\nu^\pm = -K_s^\pm \qquad J_\nu^\pm = \left(ic\rho^\pm, \mp \mathbf{J}^\pm\right) \qquad (5.11)$$

$$K_s^- = -\frac{<\hat{H}, \phi^->}{2\kappa_x \alpha_2^+}, \qquad K_s^+ = 0 \qquad (5.12)$$

where $K_s^\pm(\hat{H})$ are the conservation resources defined by a) the space dynamics of physical source $<\hat{H}, \phi^->$, and b) the time dynamics of virtual source $0 = 0^+ - 0^-$. The virtual resource is one of the important duality resources generating thermodynamics. Particularly, it is a critical source of the duality arising their next horizons of electromagnetic fields and gravitational fields, presented in the rest of the following chapters.

Therefore, the spacetime states establish twin fields of macro dynamics, which appears as a conserved current of density flux rising from the continuous symmetry and antisymmetry of spacetime interactions. These equations illustrates a principle that every differentiable symmetry of the action of a physical system has a corresponding conservation law, similar to the **Noether** first theorem[12] introduced in 1918.

CHAPTER 6

Horizon of Thermodynamics

During the formation of the second horizon at spacetime, movements of macro objects undergo interactions with and are propagated by the time fields, while events of motion objects are characterized by the space dynamics. Under the statescope of the first horizon, the spacetime dynamics of the symmetric system aggregates timestate objects to represent thermodynamics related to macro energies, statistical works, and interactive forces towards the second horizon of macroscopic variables for processes and operations characterized as a bulk system, associated with the rising temperature.

Bulk Statistics

For a bulk system of N particles, each particle is in one of three possible states: space-like $|-\rangle$, time-like $|+\rangle$, and neutral $|o\rangle$ with the energy of these states given as $E_n^{\pm}$ and E_n^o. If the bulk system has $N_n^{\pm}$ particles at non-zero charges and $N^o = N - N_n^{\pm}$ particles at neutral charge, the interruptible internal energy of the system is $E_n = N_n^{\pm} E_n^{\pm}$. The number of states $\Omega(E_n)$ of the total system of energy E_n is the number of ways to pick $N_n^{\pm}$ particles from a total of N,

$$\Omega(E) = \prod \Omega(E_n) = \prod \frac{N!}{N_n^{\pm}!(N - N_n^{\pm})!}, \quad N_n^{\pm} = \frac{E_n}{E_n^{\pm}}, \quad (6.1)$$

and the entropy is given by

$$S(E) = -k_B \log \Omega(E) = \sum_n S(E_n) = -k_B \sum_n \log \frac{N!}{(N_n^{\pm})!(N - N_n^{\pm})!} \quad (6.2)$$

where k_B is **Boltzmann** constant[13]. For large N, there is an accurate approximation to the factorials, known as the Stirling's formula[14]:

$$log\,(N!) = N \log(N) - N + \frac{1}{2} \log(2\pi N) + \Re(1/N). \quad (6.3)$$

Hence, the entropy is simplified to:

$$S(N_n^{\pm}) = -k_B N \left[\left(1 - \frac{N_n^{\pm}}{N}\right) \log\left(1 - \frac{N_n^{\pm}}{N}\right) + \frac{N_n^{\pm}}{N} \log\left(\frac{N_n^{\pm}}{N}\right) \right] \quad (6.4)$$

In general, one of the characteristics for a bulk system can be presented and measured completely by the thermal statistics of energy $k_B T$. In a bulk system with intractable energy of E_n, its temperature can be risen by its entropy S_n as the following:

$$\frac{1}{T} = \sum_n \frac{\partial S_n}{\partial E_n} = \sum_n \frac{k_B}{E_n^{\pm}} \log\left(\frac{NE_n^{\pm}}{E_n} - 1\right) \quad (6.5)$$

where the index of n is the charged or interruptible particles. Therefore, with the multiple particles of a bulk system of n particles, both of the energy of $E_n(T)$ and state probability p_n is temperature-dependent shown as the following.

$$E_n = h_n N E_n^{\pm} = \frac{N E_n^{\pm}}{e^{E_n^{\pm}/k_B T} + 1} = k_B T N_n^{\pm} \log\left(\frac{N}{N_n^{\pm}} - 1\right) \quad (6.6)$$

where $h_n = p_n$ is the horizon factor that gives rise to and emerges as the temperature T of a bulk system.

Thermodynamics

For a bulk system with the internal energy shown as above and the intractable energy of E_n, the chemical potential μ_n, rises from the following numbers of particles:

$$\mu = - \sum_n \left(\frac{\partial E_n}{\partial N_n^\pm} \right)_{S,V} = - \sum_n \left[E_n^\pm - k_B T \left(1 + e^{-E_n^\pm/k_B T} \right) \right] \qquad (6.7)$$

Its heat capacity can be given by the following definition:

$$C_V \equiv \sum_n \left(\frac{\partial E_n}{\partial T} \right)_{V,N_n^\pm} = k_B \sum_n \frac{N \left(E_n^\pm \right)^2 e^{E_n^\pm/k_B T}}{\left[k_B T \left(e^{E_n^\pm/k_B T} + 1 \right) \right]^2} \qquad (6.8)$$

The maximum heat capacity is around $T \to E^\pm/k_B$. As $T \to 0$, the specific heat exponentially drops to zero, while $T \to \infty$ drops off at a much slower pace defined by a power-law.

Consider a system with entropy $S(E, V, N_n)$ that undergoes a small change in energy, volume, and particle number $N_n^\pm$, the change in entropy is

$$dS = \frac{\partial S}{\partial E} dE + \frac{\partial S}{\partial E} \frac{\partial E}{\partial V} dV + \frac{\partial S}{\partial E} \sum_n \left(\frac{\partial E}{\partial N_n^\pm} dN_n^\pm \right) \qquad (6.9a)$$

$$dS = \frac{1}{T}\left(dE + PdV - \sum_n \mu_n dN_n^{\pm} \right), \quad P = \left(\frac{\partial E}{\partial V} \right)_T \qquad (6.9b)$$

known as the fundamental thermodynamic laws of thermodynamics of common conjugate variable pairs, developed by **Rudolf Clausius**, **William Thomson**, and **Josiah Willard Gibbs**, introduced during the period from 1850 to 1879[15].

Entropy Extrema

Furthermore, convert all of parameters to their respective densities as internal energy density $\rho_E = E/V$, thermal entropy density $\rho_s = S/V$, mole number density $\rho_{n_i} = N_i/V$, state density of $\rho_v \sim 1/V$. The equation of (6.9) becomes the entropy relationship in terms of their following density:

$$S_\rho = -\int \rho_v d\Gamma = \int \frac{d\rho_E - Td\rho_s - \sum_i \mu_i d\rho_{n_i}}{T\rho_s + \sum_i \mu_i \rho_{n_i} - (P + \rho_E)} d\Gamma \qquad (6.10)$$

Satisfying entropy equilibrium at extrema results in the general density equations of the thermodynamic fields:

$$d\rho_E^- = Td\rho_s^- + \sum_i \mu_i d\rho_{n_i}^- \qquad (6.11)$$

$$P + \rho_E^+ = T\rho_s^+ + \sum_i \mu_i \rho_{n_i}^+ \qquad (6.12)$$

The first equation of (6.11) indicates that physical *Space Entropy* increases towards maximum in physical disorder, so that the dynamics of the internal energy are the interactive fields of thermal entropy and chemical reactions as they influence substance molarity. The second equation (6.*12*) indicates that virtual *Time Entropy* decreases towards minimum in physical order, so that both external forces and internal energy hold balanced macroscopic fields in a bulk system.

CHAPTER 7

Horizon of Electromagnetic Fields

The word electromagnetism is defined in terms of the electromagnetic fields, including both electricity and magnetism as a duality of manifestations of the Indivisible phenomenon.

The electromagnetic fields play a major role in determining a part of internal transformation of most objects encountered in daily life. Ordinary matter takes its form as a result of intermolecular communicating between individual molecules. For example, carrying intrinsic messages, electrons are traveling over electromagnetic fields and bounding at orbitals around atomic nuclei to form atoms, which are the building blocks of molecules. This governs the processes involved in chemistry horizon, arising from interactions between the electrons of neighboring atoms, which are in turn determined by the interactions among electromagnetic fields and the momentum of the electrons.

At high energy horizon of macroscopic density, the electromagnetic forces are unified under the duality of spacetime dynamic fields, where electromagnetism is considered as one of the fundamental forces in the regime of macroscopic horizon.

The spacetime dynamic field can be viewed as the combination of the electric-magnetic and space-time fields. The electric field is contributed primarily by spacial charges, while the magnetic field primarily by moving charges, the spacetime duality of resources for the fields. The way in which charges and currents interact with the electromagnetic field is described by the 2x2 mixture of electric-magnetic and space-time dynamics, known as the Maxwell's equations and the Lorentz forces.

Electromagnetic Fields

In macroscopic scope, a conserved current at the third horizon implies a conserved charge $Q^\pm = \int d^3x J^\pm$ of the given states, which creates physical electric fields **D**, a vector field, and virtual magnetic fields **B**, a pseudo-vector field. In free space of a macro system, the **D** field rises during the motion of physical objects while **B** field rises simultaneously from the inevitable duality of virtual objects:

$$\rho^- = \nabla \cdot \mathbf{D} \tag{7.1a}$$

$$\rho^+ = \nabla \cdot \mathbf{B} \tag{7.1b}$$

Combined with spacetime antisymmetry, a duality of space and time fields of the additional virtual sources are undetectable in physical measurements:

$$-K_s^- + 0^+ \Rightarrow -\nabla \cdot \mathbf{J}_s^+ + \kappa_B \nabla \cdot (\nabla \times \mathbf{B}) \tag{7.2a}$$

$$-K_s^+ - 0^- \Rightarrow -\nabla \cdot \mathbf{J}_s^- - \kappa_D \nabla \cdot (\nabla \times \mathbf{D}) \tag{7.2b}$$

where $\mathbf{J}_s^\mp$ are space and time sources of current densities. Substitute equation (5.10) by equations (7.1) and (7.2), a general relation of the space and time conserved currents gives rise to the macroscopic space and time fields of **D** and **B**:

$$\nabla \cdot (ic\frac{\partial \mathbf{D}}{\partial x_0} + \mathbf{J}^-) = -\nabla \cdot (\mathbf{J}_s^+ - \kappa_B \nabla \times \mathbf{B}) \qquad (7.3a)$$

$$\nabla \cdot (ic\frac{\partial \mathbf{B}}{\partial x_0} - \mathbf{J}^+) = -\nabla \cdot (\mathbf{J}_s^- + \kappa_D \nabla \times \mathbf{D}) \qquad (7.3b)$$

These equations give rise to electromagnetic fields generated by the macro densities of both physical and virtual densities and current distributions, respectively, shown by the following differential equations:

$$\rho_q = \nabla \cdot \mathbf{D} \qquad\qquad : \rho_q = \rho^- \qquad\qquad (7.4a)$$

$$ic\frac{\partial \mathbf{D}}{\partial x_0} + \mathbf{J} = \kappa_B \nabla \times \mathbf{B} \quad : \mathbf{J} = \mathbf{J}^- + \mathbf{J}_s^+, \quad \mu_r \kappa_B = 1 \qquad (7.4b)$$

$$\rho^+ = \nabla \cdot \mathbf{B} = 0 \qquad\qquad : J_0^+ = 0 \qquad\qquad (7.4c)$$

$$ic\frac{\partial \mathbf{B}}{\partial x_0} = -\kappa_D \nabla \times \mathbf{D} \quad : \mathbf{J}^+ - \mathbf{J}_s^- = 0, \quad \kappa_D = 1 \qquad (7.4d)$$

known as the **Maxwell** equations or electromagnetic fields[16], introduced in 1861. Experimentally, the equation (7.4c) is known as **Gauss** law[17], introduced in 1835, for magnetic fields that total magnetic flux rises from spacetime interactions, while the equation (7.4a) is known as the electric flux sourced by the charges. The equation (7.4d) is known as the **Faraday** Law[18], introduced in 1831, which states a dynamic magnetic field gives rise to a circulating electric field. The equation (7.4b) is known as the **Ampere** Law[19], introduced in 1826, which states a electric flow of physical current gives rise to a magnetic virtual field that circles the wire, and that a dynamic electric field gives rise to a magnetic field encircling the electric fields.

Covariant Expression

The covariant formulation of electromagnetic fields yields the field tensor of physical $\check{F}_{\mu\nu}$ and virtual $\hat{F}_{\mu\nu}$. Each has a twin pair of its antisymmetry tensors $\check{F}'_{\mu\nu}$ and $\hat{F}'_{\mu\nu}$, respectively, defined as pseudo-tensors, a transformation that can be expressed as a proper rotation followed by sign reflection. Equivalent to *Maxwell* equations of (7.4a) and (7.4b), the electromagnetic fields in a covariant formulation and components have the following net expression:

$$\partial_\mu \check{F}_{\mu\nu} = J^\nu \qquad : J^\nu \in \left(ic\rho_q, \mathbf{J} \right) \tag{7.5a}$$

$$\check{F}_{\mu\nu} = \begin{pmatrix} 0 & -icB_1 & -icB_2 & -icB_3 \\ icB_1 & 0 & D_3 & -D_2 \\ icB_2 & -D_3 & 0 & D_1 \\ icB_3 & D_2 & -D_1 & 0 \end{pmatrix} \tag{7.5b}$$

where the three spatial components are the electric field $\mathbf{E}$ and magnetic field $\mathbf{B}$, respectively. Likewise, the equations (7.4c) and (7.4d) have the following covariant formulation:

$$\partial_\mu \hat{F}_{\mu\nu} = J^o \qquad : J^o_\nu \in \left(ic\rho^+, \mathbf{J}^+ - \mathbf{J}^-_s \right) \Rightarrow 0^+ \tag{7.6a}$$

$$\hat{F}_{\mu\nu} = \begin{pmatrix} 0 & -icD_1 & -icD_2 & -icD_3 \\ icD_1 & 0 & -H_3 & H_2 \\ icD_2 & H_3 & 0 & -H_1 \\ icD_3 & -H_2 & H_1 & 0 \end{pmatrix} \qquad\qquad (7.6b)$$

where the three spatial components are the electric displacement field $\mathbf{D} = \varepsilon_r \mathbf{E}$ and magnetic intensity field $\mathbf{H} = \mathbf{B}/\mu_r$, respectively.

Physical Potential Dynamics

Furthermore, a covariance manifest of electromagnetic fields $\check{F}^{\mu\nu}$ can be made by a 4-potential vector of A_ν, with which a 4-derivative vector can perform their operations as an electromagnetic tensor, shown by the following:

$$\check{F}_{\mu\nu} = \partial_\mu A_\nu - \partial_\nu A_\mu : \quad A_\nu = \{ic\Phi, \mathbf{A}\} \qquad\qquad (7.7)$$

where Φ is a scalar potential and $\mathbf{A}$ is a vector potential. Obviously, we derives both of the electric and magnetic potentials as the following relationship:

$$\mathbf{B} = \nabla \times \mathbf{A} \qquad \mathbf{E} = -\nabla\Phi - \frac{\partial \mathbf{A}}{\partial t} \qquad\qquad (7.8)$$

Therefore, the potentials rise from the electromagnetic fields.

In macroscopic spacetime, movements of a macro object can be expressed by characteristics of 4-velocity U^μ as given by equation

$$U^\mu \in ic\frac{\partial^\mu}{\partial x_0}x^\mu = (ic, \mathbf{u}) \quad : U^\mu U^\nu = c^2 \tag{7.9}$$

Because the electromagnetic force $\mathbf{F}_q$ acting on a macroscopic object at charge q is defined by

$$F_q^\nu = ic\frac{\partial P_q^\nu}{\partial x_0} = qU^\mu F_{\mu\nu} \tag{7.10}$$

this formula arranges itself into a vector equation of:

$$\mathbf{F}_q = \frac{\partial \mathbf{P}_q}{\partial t} = q\,(\mathbf{E} + \mathbf{u} \times \mathbf{B}) = q(-\nabla\Phi + \frac{\partial \mathbf{A}}{\partial t} + \mathbf{u} \times \nabla \times \mathbf{A}) \tag{7.11}$$

known as **Lorentz** force[20], introduced by **Oliver Heaviside**[21] in 1889, that rise from electromagnetic fields or their associated potentials.

Summary

In conclusion, the physical space dynamics of electric fields has as its opponent of virtual time dynamics of magnetic fields, which form and give rise to a macroscopically coherent fabric of our natural horizons, exhibited throughout all physical existence.

CHAPTER 8

Horizon of Gravitational Fields

In macroscopic, the dynamics of $\mathbf{J}_s^{\pm}$ are the interactive currents from external sources while $\mathbf{J}^{\pm}$ are the induced currents from its own motion. Under space primacy, one of the twins forms the electromagnetic current of $\mathbf{J} = \mathbf{J}^- + \mathbf{J}_s^+$, shown by equation (7.4b). Meanwhile, under time primacy, the other twin forms the virtual current equation (7.4d) of $\mathbf{J} = \mathbf{J}^+ - \mathbf{J}_s^- = 0$, the time dynamics of virtual sources.

Gravitational Fields

The virtual resource of $0 = 0^+ - 0^-$ is a critical source of the duality arising the horizon of gravitational fields, presented by gravitational tensors. From equation (5.11), this can be further defined by the following:

$$\mathbf{J}^+ = 0^+ \Rightarrow \partial_\mu \hat{G}^{\mu\nu} \quad : \hat{G}^{\mu\nu} = \hat{G}^{\nu\mu} \tag{8.1}$$

$$\mathbf{J}_s^- = 0^- \Rightarrow \partial_\mu \check{G}^{\mu\nu} \quad : \check{G}^{\mu\nu} = \check{G}^{\nu\mu} \tag{8.2}$$

$$\partial_\mu \partial^\nu \mathbf{F}_{\mu\nu} = \partial_\mu \hat{G}^{\mu\nu} - \partial_\mu \check{G}^{\mu\nu} = \mathbf{0} \tag{8.3}$$

The tensor $\hat{G}^{\mu\nu}$ is induced from its own time nature 0^+ of dynamics in both virtual and physical worlds, called movement tensor, presenting a time dynamics of curvature tensor that determines the motion degree of spacetime to which a physical object tends to converge or diverge in virtual time. The tensor $\check{G}^{\mu\nu}$ is an interactive source of objects, called stress tensor, that sources a space nature 0^- of energy tensor $T^{\mu\nu}$ and generalizes to the relative potential of forces with which an object tends to have attraction or restraint in space.

Rising from the time primary of density current, the $\check{G}$ measures the energy and momentum contained the twin tensors, and the $\hat{G}$ measures the virtual curvature of spacetime, $R_{\mu\nu}$,

$$\hat{G} = R \qquad : \hat{G}^{\mu\nu} = R^{\mu\nu} \tag{8.4}$$

where $R_{\mu\nu}$ is known as the **Ricci** tensor[22] introduced in 1880s.

Because the space tensor $\check{G}$ is a symmetric and divergence-less tensor, it is proportional to the stress energy tensor **T**, shown by the following:

$$\check{G} = h_g T \qquad : \check{G}^{\mu\nu} = h_g T^{\mu\nu}, \quad \partial_\mu T_{\mu\nu} = 0, \quad T_{\mu\nu} = T_{\nu\mu} \tag{8.5}$$

where h_g is a horizon constant. A stress energy tensor **T** is a source of the gravitational fields. Matters of mass density under physical primacy interact with its twin of space fields under virtual primacy, bending an object curvature, and curving its matter's movement.

As a physical massive body extends into the space around itself, it encompasses all energy and momentum of matter fields, acting as a source of gravity, and producing the virtual force on other massive objects.

$$h_g T^{\mu\nu} = R^{\mu\nu} \tag{8.6}$$

This derives a general equation of the twin tensors, an stress-energy tensor sourced from physical mass of space objects and a curvature-momentum tensor sourced from virtual massless of time instances. Together, they rise from spacetime movements.

Virtual Potential Dynamics

For a common macro system in spacetime, the stress-energy tensors of physical interactions can be expressed as a time independent matrix of the following form:

$$T = \begin{pmatrix} c^2\rho & 0 & 0 & 0 \\ 0 & & & \\ 0 & & \delta_{ij}\sigma & \\ 0 & & & \end{pmatrix}, \quad \sigma = \begin{pmatrix} \sigma_{xx} & \sigma_{xy} & \sigma_{xz} \\ \sigma_{yx} & \sigma_{yy} & \sigma_{yz} \\ \sigma_{zx} & \sigma_{zy} & \sigma_{zz} \end{pmatrix} \tag{8.7}$$

where ρ_m is the mass destiny, σ_{ij} is the **Cauchy** stress tensor[23], introduced in 1827. Under a homogenous environment, the tensor is reduced to

$$\sigma_{ij} = p_0\delta_j^i \qquad : i, j \in (1,2,3) \tag{8.8}$$

where p_0 is a constant pressure. In a covariant form, this tensor is equivalent to the forms:

$$T^{\mu\nu} = \left(\rho_m + \frac{p_0}{c^2}\right)U^\mu U^\nu - p_0 g^{\mu\nu} \quad : T^{\mu\nu} \in \left(c^2\rho_m, \sigma_{ij}\right) \tag{8.9}$$

From these relationships, we have the following equations:

$$\hat{G}^{00} = R^{00} = -\Gamma_{00}^\mu = -\frac{2}{c^2}\nabla \cdot \mathbf{g} \quad : \mathbf{g} \equiv g_{00} = -\nabla\Phi \tag{8.10a}$$

$$\check{G}^{00} = h_g T^{00} = h_g c^2 \rho_m \tag{8.10b}$$

where using kinetic operator of $\mathbf{g} \Rightarrow -\nabla\Phi$ for a point object, the macro entity of the spacetime duality illustrates cyclic of virtual and physical nature within a point object, which poses a conservative force such that its field has divergence of non-rotational or zero curl: $\nabla \times \mathbf{g}\left(x_v\right) = 0$.

Therefore, in an environment of homogeneity and time independence, the mass density is the source of potential fields Φ in **Newtonian** gravity[24] with the universal gravitational constant G_0 as the following:

$$\nabla^2\Phi = 4\pi G_0 \rho_m \quad : h_g = \frac{8\pi G_0}{c^4} \tag{8.11}$$

known as the **Poisson** equation[25] for gravity, introduced in 1813. Alternatively, the above equation becomes the following:

$$\nabla \cdot \mathbf{g}\left(x_v\right) = -4\pi G_0 \rho_m \tag{8.12}$$

known as **Gauss's law**[17] for gravity. For a point mass, the gravitational field $\mathbf{g(r)}$ is given by the following:

$$\mathbf{g(r)} = -m_0 G_0 \frac{\mathbf{r}}{r^2} \quad : \oint \rho_m d\Gamma = -\frac{m_0 \mathbf{r}}{4\pi r^2} \tag{8.13}$$

This gravitational field is a virtual dynamics, which, in mathematics, is presented as a vector of the gravitational force applied on an object in any given point in space with per unit mass of m as the following:

$$F(\mathbf{r}) = m\mathbf{g(r)} = -mm_0 G_0 \frac{\mathbf{r}}{r^2} \tag{8.14}$$

known as **Newton**'s law[26], introduced in 1687. As we know, $g(r)$ is also equivalent to the gravitational acceleration at that point.

A spacetime with a gravitational singularity presents the nature of event horizons in macroscopic regime extended from one horizon to the other, within each of which there exist the relatively significant theorems. This singularity does represent the initial state at its macroscopic horizon. It might not mean there are the singularities neither at the universe, nor at the beginning of the Big Bang[30], nor at the inside black hole, which are known as the hypothetical singularities.

General Relativity

Generally, the right-side of equations (8.1) can be derived using a constant, $R/2$, instead of zero, 0. Let us rewrite the equations with the following notations:

$$J^+ = 0^+ \Rightarrow \partial_\mu G_{\mu\nu} \quad : G_{\mu\nu} = G_{\nu\mu} \qquad (8.15)$$

$$J_s^- = 0^- \Rightarrow \partial_\mu R_{\mu\nu} \quad : R_{\mu\nu} = R_{\nu\mu} \qquad (8.16)$$

The equation (8.1) becomes the following non-zero tensor forms:

$$\partial_\mu \partial^\nu \hat{F}_{\mu\nu} = \partial_\mu G_{\mu\nu} - \partial_\mu R_{\mu\nu} = -\partial_\mu (\frac{1}{2} R g_{\mu\nu}) = 0 \qquad (8.17)$$

$$G_{\mu\nu} \equiv h_g T_{\mu\nu} = R_{\mu\nu} - \frac{1}{2} R g_{\mu\nu} \quad : \partial_\mu G_{\mu\nu} = \partial_\mu R_{\mu\nu}, \quad \partial_\mu g_{\mu\nu} = 0 \qquad (8.18)$$

where the constant R is a *Ricci* curvature scalar[27], and $g_{\mu\nu}$ is a metric tensor in a four-dimensional manifold. By introducing the horizon constant $h_g = 8\pi G_0/c^4$ of equation (8.11), the equation (8.18) becomes the following

$$\frac{8\pi G_0}{c^4}T_{\mu\nu} = R_{\mu\nu} - \frac{1}{2}Rg_{\mu\nu} \qquad\qquad (8.19)$$

known as the **Einstein** field equation[28], introduced in November 1915.

Assuming a static universe, Albert Einstein found that the constant R in equation (8.19) may collapse the universe and require a hypothetical repulsive force for the balancing. Therefore, he introduced a "cosmological constant" of Λ to rewrite the above equation with $R \dashrightarrow R - 2\Lambda$, into the following:

$$\frac{8\pi G_0}{c^4}T_{\mu\nu} = R_{\mu\nu} - \frac{1}{2}Rg_{\mu\nu} + \Lambda g_{\mu\nu} \qquad\qquad (8.20)$$

In the 1920s, astronomer **Edwin Hubble**[29] discovered that the distances to faraway galaxies were strongly correlated with their redshifts. In accordance with that universe was expanding, Einstein discarded the cosmological constant as an unnecessary fudge factor and called it the "biggest blunder of my life".

Without using the interwoven continuum of energy state and horizon fields, although the equations of (8.19) and (8.20) were derived by Einstein, the original meaning of general relativity was philosophically difficulty to understand even after his research over next thirty-five years, declared in the following statements:

✦ *"The general theory of relativity is as yet incomplete insofar as it has been able to apply the general principle of relativity satisfactorily only to gravitational fields, but not to the total field. We do not yet know with certainty, by what mathematical mechanism the total field in space is to be described and what the general invariant laws are to which this total field is subject. One thing, however, seems certain: namely, that the general principle of relativity will prove a necessary and effective tool for the solution of the problem of the total field." - Albert Einstein, "The theory of relativity", 1949*

✦ *"According to general relativity, the concept of space detached from any physical content does not exist."..."all attempts to obtain a deeper knowledge of the foundations of physics seem doomed to me unless the basic concepts are in accordance with general relativity from the beginning. This ... forces us to apply free speculation to a much greater extent than is presently assumed by most physicists." - Albert Einstein, "On the generalized theory of gravitation" April 1950*

This marks the centenary of that discovery, which hands up with the philosophical ambiguity.

Summary

Gravity is a natural phenomenon of space-time duality rising from virtual dynamics under time primacy (or traditionally named as the curvature of spacetime). Its twin is the physical dynamics under space primacy, giving rise to the electromagnetic dynamics. As a consequence, gravity is a force in the regime of macroscopic horizon, caused by the uneven distribution of energy and time dilation.

Mathematically, the general relativity equation of (8.19) contains the *Newtonian* gravity when the constant R of a *Ricci* curvature scalar is normally neglectable. In reality, it is critical to keep the equation (8.17) in mind:

$$\partial_\mu g_{\mu\upsilon} = \partial_\mu G_{\mu\nu} - \partial_\mu R_{\mu\nu} = 0 \qquad\qquad (8.21)$$

Our universe has the conservations that the physically observable is expanding while the virtually unobservable can be opposite in balancing. This leaflet might be helpful in answering the collapse of universe, assuming all observable regions of the universe are receding from all others. The continuous expansion implies that there is denser and warmer as a duality in reverse.

CHAPTER 9

Conclusions

The topological duality and evolutionary processes of spacetime horizons derive a complete picture of the principal equations, important assumptions, and essential laws, discovered and described by classical physics since three hundred thirty years ago, start from Newton's mechanics of 1687, to Maxwell's equations of 1981, to Einstein's relativity of 1915, and to quantum physics of 20th-century. It prompts the unified horizon topology of virtual and physical dynamics, from nuclear particles[31], to quantum physics, to thermodynamic bulk densities, to electromagnetic and gravitational fields, to macroscopic mechanics, to look back to the future: Virtumanity - Duality of Virtual and Physical Reality.

References

[1] wikipedia. https://en.wikipedia.org/wiki/Lagrangian_mechanics

[2] wikipedia. https://en.wikipedia.org/wiki/Euler–Lagrange_equation

[3] Nathalie, B., Hartshorne, G., Cavilla, J., Taylor, J., Gardner, J., Wilmut, I., Meehan, R., and Young, L., Non-conservation of mammalian preimplantation methylation dynamics, Volume 14, Issue 7, pR266–R267, 6 April 2004 [3] wikipedia. https://en.wikipedia.org/wiki/Taylor_series

[4] wikipedia .https://en.wikipedia.org/wiki/Hamiltonian_(quantum_mechanics)

[5] wikipedia. https://en.wikipedia.org/wiki/Schrödinger_equation

[6] wikipedia. https://en.wikipedia.org/wiki/Planck_constant

[7] wikipedia. https://en.wikipedia.org/wiki/Mass–energy_equivalence

[8] wikipedia. https://en.wikipedia.org/wiki/Dirac_equation

[9] wikipedia. https://en.wikipedia.org/wiki/Klein–Gordon_equation

[10] wikipedia. https://en.wikipedia.org/wiki/Kelvin

[11] wikipedia. https://en.wikipedia.org/wiki/Heisenberg_picture

[12] wikipedia. https://en.wikipedia.org/wiki/Noether%27s_theorem

[13] wikipedia. https://en.wikipedia.org/wiki/Boltzmann_constant

[14] wikipedia. https://en.wikipedia.org/wiki/Stirling%27s_approximation

[15] wikipedia. https://en.wikipedia.org/wiki/Thermodynamics

[16] wikipedia. https://en.wikipedia.org/wiki/Maxwell%27s_equations

[17] wikipedia. https://en.wikipedia.org/wiki/Gauss%27s_law

[18] wikipedia. https://en.wikipedia.org/wiki/Faraday%27s_law_of_induction

[19] wikipedia. https://en.wikipedia.org/wiki/Ampère%27s_circuital_law

[20] wikipedia. https://en.wikipedia.org/wiki/Lorentz_force

[21] wikipedia. https://en.wikipedia.org/wiki/Oliver_Heaviside

[22] wikipedia. https://en.wikipedia.org/wiki/Ricci_curvature

[23] wikipedia. https://en.wikipedia.org/wiki/Cauchy_stress_tensor

[24] wikipedia. https://en.wikipedia.org/wiki/Newton%27s_law_of_universal_gravitation

[25] wikipedia. https://en.wikipedia.org/wiki/Poisson%27s_equation

[26] wikipedia. https://en.wikipedia.org/wiki/Newton%27s_laws_of_motion

[27] wikipedia. https://en.wikipedia.org/wiki/Scalar_curvature

[28] Einstein, Albert. Relativity: The Special and General Theory, 1916. http://www.emu.dk/sites/default/files/relativity.pdf

[29]: https://en.wikipedia.org/wiki/Edwin_Hubble

[30]: https://en.wikipedia.org/wiki/Big_Bang

[31] Xu, C. Wei. Theory of Physical Cosmology - Universe Particles, March 25, 2015, ISBN: 9-78099-0520610. iBook: https://itunes.apple.com/us/book/id1044006505

www.ingramcontent.com/pod-product-compliance
Lightning Source LLC
Chambersburg PA
CBHW060419190526
45169CB00002B/964